NUREG-0561, Rev. 2

I0493822

Physical Protection of Shipments of Irradiated Reactor Fuel

Final Report

Office of Nuclear Security and Incident Response

AVAILABILITY OF REFERENCE MATERIALS
IN NRC PUBLICATIONS

United States Nuclear Regulatory Commission

Protecting People and the Environment

NUREG-0561, Rev. 2

Physical Protection of Shipments of Irradiated Reactor Fuel

Final Report

Manuscript Completed: January 2013
Date Published: April 2013

Prepared by:
Pacific Northwest National Laboratory
P.O. Box 999
Richland, WA 99352

and

Division of Security Policy
Office of Nuclear Security and Incident Response
U.S. Nuclear Regulatory Commission
Washington, DC 20555-0001

ABSTRACT

This guidance document sets forth means, methods and procedures that the staff of the U.S. Nuclear Regulatory Commission (NRC) considers acceptable for satisfying the requirements for the physical protection of spent nuclear fuel (SNF) during transportation by road, rail and water, and for satisfying the requirements for background investigations of individuals granted unescorted access to SNF during transportation. Chapter 1 discusses the regulatory basis and definitions applicable to shipments of SNF. Chapter 2 corresponds to the general requirements for the physical protection of SNF while in transit. These requirements apply to all SNF shipments regardless of the mode of transportation used for a particular shipment. Chapters 3, 4, and 5 discuss additional requirements specific to each particular transport mode—road, rail, or U.S. waters. Chapter 6 discusses the requirements for background investigations of individuals with unescorted access to SNF in transit.

Paperwork Reduction Act Statement

Public Protection Notification

CONTENTS

1. INTRODUCTION

This chapter discusses the regulatory authority for the U.S. Nuclear Regulatory Commission (NRC) requirements for physical protection of and access to spent nuclear fuel (SNF) shipments. The chapter also includes definitions of terms used in this guidance and a discussion of the organization of this NUREG.

1.1. Regulatory Authority

The guidance in this NUREG pertains to the physical protection of irradiated reactor fuel during transportation and to background investigations of individuals granted unescorted access to irradiated reactor fuel during transportation, as required by U.S. Nuclear Regulatory Commission (NRC) regulations in Title 10 of the *Code of Federal Regulations* (10 CFR) 73.37, "Requirements for Physical Protection of Irradiated Reactor Fuel in Transit," and 10 CFR 73.38, "Personnel Access Authorization Requirements for Irradiated Reactor Fuel in Transit."

The requirements of 10 CFR 73.37 apply to each licensee who transports, or delivers to a carrier for transport, in a single shipment, a quantity of irradiated reactor fuel in excess of 100 grams (0.22 lbs) in net weight of irradiated fuel, exclusive of cladding or other structural or packaging material, which has a total external radiation dose rate in excess of 1 Gy (100 rad) per hour at a distance of 1 meter (3.3 feet) from any accessible surface without intervening shielding.

For the purposes of 10 CFR 73.37, 10 CFR 73.38, and this NUREG, the terms "irradiated reactor fuel" and "spent nuclear fuel" are considered synonymous and are used interchangeably. In addition, references to "the staff" in this NUREG means the NRC staff, unless otherwise noted.

These regulations apply to SNF from either power or nonpower reactors that is contained in a domestic shipment or the domestic portions of import or export shipments (i.e., while the shipment is within U.S. territory or U.S. territorial waters). The 100 gram threshold quantity is intended to apply to the combination of uranium and plutonium compounds and associated fission products generated during irradiation. The weight of the fuel cladding or other structural or packaging material associated with the fuel should not be included in determining whether the quantity of SNF in the given shipment needs to be protected under the requirements of 10 CFR 73.37.

The objective of the physical protection system for shipments of such material is to minimize the potential for theft, diversion, or radiological sabotage of SNF shipments and to facilitate the location and recovery of SNF shipments that may have come under the control of unauthorized individuals. The NRC is cognizant that radiological sabotage could result in economic consequences and social disruptions. The staff believes these impacts will be minimized if the security requirements in 10 CFR Part 73 are effectively implemented. As noted in the Final Rule, the Commission did not change the definition of radiological sabotage in 10 CFR 73.2. The term radiological sabotage as used in this guidance document has the same meaning as the definition in 10 CFR 73.2.

The licensee shall establish a physical protection system for SNF in transit that includes armed escorts and a movement control center staffed and equipped to monitor and control shipments, communicate with local law enforcement authorities (LLEA), and respond to normal conditions

and security contingencies. In accordance with the performance objectives of
10 CFR 73.37(a)(2), the physical protection system shall do the following:

- Provide early detection and assessment of attempts to gain unauthorized access to, or control over, SNF shipments.

- Delay and impede attempts at theft, diversion, or radiological sabotage of SNF shipments.

- Provide notification to the appropriate response forces of any attempts at theft, diversion, or radiological sabotage of SNF shipments.

Before granting an individual unescorted access to SNF in transit, the licensee shall determine if the individual is trustworthy and reliable by completing a background investigation of the individual in accordance with 10 CFR 73.38. The requirements of 10 CFR 73.38 apply to vehicle operators, escorts, train crews, and any other individuals accompanying the shipment during transport, as well as movement control center personnel. Fingerprinting and the other investigative elements in 10 CFR 73.38 do not apply to Federal, State, local, and Tribal[1] law enforcement personnel, nor do they apply to other persons identified in 10 CFR 73.59, "Relief from Fingerprinting, Identification and Criminal History Record Checks for Designated Categories of Individuals," as well as other categories of persons, as specified in 10 CFR 73.38(c)(2).

1.2. Organization of NUREG-0561

The guidance contained in this document is organized to correspond to applicable sections of 10 CFR 73.37 and 10 CFR 73.38. Chapter 1 describes the regulatory basis and definitions applicable to shipments of SNF. Chapter 2 corresponds to the general requirements contained in 10 CFR 73.37(b) and 10 CFR 73.37(f). These requirements apply to all SNF shipments regardless of the mode of transportation used for a particular shipment. Chapters 3, 4, and 5 correspond to 10 CFR 73.37(c), 10 CFR 73.37(d), and 10 CFR 73.37(e), respectively, which contain additional requirements specific to each particular transport mode—road, rail, or U.S. waters. Chapter 6 corresponds to 10 CFR 73.38, which addresses requirements for background investigations of individuals with unescorted access to SNF in transit.

In addition to the specific physical protection requirements in 10 CFR 73.37(b), 10 CFR 73.37(c), 10 CFR 73.37(d), 10 CFR 73.37(e), and 10 CFR 73.37(f), the regulation includes performance objectives in 10 CFR 73.37(a). These performance objectives are not intended to be implemented as specific requirements, but rather are general statements of the Commission's intent. These objectives characterize and further define the general level of protection that physical protection systems designed to satisfy the requirements of the regulation are expected to provide. In this document, discussions about the requirements in 10 CFR 73.37(b) through (f) have been developed to be consistent with and to support the performance objectives in 10 CFR 73.37(a) to the extent that each objective applies to a specific element of the physical protection system.

---

[1] After June 11, 2013, licensees must comply with the requirements in this document pertaining to participating Tribes, Tribal law enforcement personnel, Tribal officials, Tribal official's designees and Tribal reservations (77 FR 34194; June 11, 2012).

1.3. Definitions

The following definitions apply to terms used in this guidance document. These definitions shall apply irrespective of the mode of transport, except as otherwise noted. Terms used in this guidance not defined below shall have the same meaning as defined in 10 CFR 73.2, "Definitions."

Escort: A person with similar duties to that of an armed escort, who may or may not be armed.

Local law enforcement agency (or authority) (LLEA): Any State, Tribal, county, or municipal agency that has law enforcement authority within the locality or jurisdiction through which the shipment may pass. The term is usually limited to the particular law enforcement agencies responsible for responding to calls for assistance by escorts, such as county or municipal police forces, port authority police, railroad police, or the highway patrol. In some instances involving waterborne shipments, the U.S. Coast Guard or another Federal agency also may be considered as the LLEA, depending on whether or not they are relied on to provide a response to a request for assistance.

LLEA radio communications: The radio communications system that the LLEA normally uses to communicate with its mobile units.

Monitor: Capability to observe and detect unauthorized access.

Participating Tribe: An Indian Tribe, as defined in 10 CFR 73.2, which has notified the NRC that it would like to receive advance notification of any shipment of SNF that passes within or across the Tribal reservation and that has certified to the NRC that any Safeguards Information (SGI) it receives will be appropriately protected.

Radiological sabotage: As defined in 10 CFR 73.2, this term refers to any deliberate act directed against a plant or transport in which an activity licensed under the regulations in this chapter is conducted, or against a component of such a plant or transport that could directly or indirectly endanger the public health and safety by exposure to radiation.

Safe haven: A readily recognizable and accessible site at which security is present or from which, in the event of an emergency, the transport crew can notify and wait for LLEA response. A safe haven is an area that can provide prudent security measures when a shipment is removed from transit because of an elevated threat condition.

Telemetric monitoring system: A data transfer system that captures information by instrumentation or measuring devices about the location and status of a transport vehicle or package between the departure and destination location. The U.S. Department of Energy's TRANSCOM system is one example of a telemetric monitoring system.

Transport vehicle or shipment vehicle: For a road shipment, an integrated-unit type of truck or combination tractor-trailer that bears an SNF shipment; for a rail shipment, the shipment vehicle or shipment car is the rail car containing the SNF package. In 10 CFR 73.37, it is written in terms of a single transport vehicle or shipment car being used for a given SNF shipment. However, it is recognized that more than one such vehicle may be used in a single shipment. In this case, all provisions of the regulation should be interpreted to apply to all of the transport vehicles as a group, unless noted otherwise in the guidance.

2. GENERAL REQUIREMENTS

This chapter provides guidance on the U.S. Nuclear Regulatory Commission (NRC) requirements for physical protection of spent nuclear fuel (SNF) shipments made by road, rail or water. Requirements discussed in this chapter apply to all SNF shipments regardless of the mode of transport, unless otherwise noted. Chapters 3, 4 and 5 provide additional guidance on requirements specific to physical protection of road, rail and waterborne SNF shipments, respectfully.

2.1. NRC Approval of SNF Shipment Routes

In 10 CFR 73.37(b)(1)(vi), the NRC requires the licensee to obtain NRC approval for the planned road and rail routes over which SNF is to be shipped and for any U.S. ports where a vessel carrying an SNF shipment is scheduled to dock. Whenever possible, the licensee should request approval of two routes (i.e., a primary and an alternative) when transporting SNF by road or rail.

Section 73.37(b)(1)(vi) requires that the licensee obtain the route approval in writing from the NRC before the 10-day advanced notification requirement in 10 CFR 73.72(a)(2). Licensees should submit applications for route approvals at least 6 months before the planned date of the shipment to allow the staff adequate time to review and approve the requested route. Requests to approve a route that do not comply with this requirement will be handled on a case-by-case basis.

In 10 CFR 73.37(b(1)(iv), the NRC requires that routes used for shipping SNF comply with applicable requirements of the U.S. Department of Transportation (DOT) regulations in Title 49 of the Code of Federal Regulations (49 CFR), in particular, those identified in 10 CFR 71.5, "Transportation of Licensed Material." For shipments by road, DOT requirements for the security of radioactive material shipments are found in Subpart I "Safety and Security Plans," to 49 CFR Part 172, "Hazardous Materials Table, Special Provisions, Hazardous Materials Communications, Emergency Response Information, and Training Requirements," and Subpart D, "Routing of Class 7 (Radioactive) Materials," to 49 CFR Part 397 "Transportation of Hazardous Materials; Driving and Parking Rules." The licensee should document that the highway route complies with DOT routing requirements of Subpart D to 49 CFR Part 397, or that DOT approval has been obtained if it does not. For shipments by rail, DOT requirements for the security of radioactive material shipments are found in 49 CFR Part 172, 49 CFR Part 174, "Carriage By Rail," and 49 CFR Part 209, "Railroad Safety Enforcement Procedures."

The application for an NRC route approval shall include the following information about the planned shipping campaign and route as required by Section 73.37(b)(1)(vi):

- Shipper, consignee (receiver), carriers, transfer points and modes of shipment

- Statement of shipment security arrangements, including, if applicable, points where armed escorts transfer responsibility for the shipment

- Location of safe havens that have been coordinated with the appropriate States(s) for road shipments

The licensee should also include the following in the route approval application:

- Description of the cargo, including type of SNF, quantity of irradiated fuel elements, and the radiation level at 1 meter unshielded (e.g., reactor type, number of elements, and confirmation that the absorbed dose rate is greater than 1 gray per hour)

- Package identification (e.g., package design, package capacity). (Note: specific package identification parameters (e.g., serial number) may be submitted later if unknown when the route approval request is submitted to the NRC)

- Loaded weight of road transport vehicle or shipment rail car; for waterborne shipments, weight of loaded package or cargo containers

- Information about the planned shipping campaign schedule, including the following:
  – number of shipments proposed in a series of shipments
  – approximate duration of each shipment from point of origin to destination
  – anticipated date(s) of shipment(s)

- Justification for the staff's priority review of the proposed route if the shipment is planned for less than 60 days from the date of the approval request

- Description of the route to be used, including route plan and mileage, safe havens along the route, summary map of the route, detailed maps of the route, contact information along the route, communications capability along the route, and confirmation that a route inspection was performed (the description of the route selection criteria and application format is discussed in further detail in Sections 2.1.1 and 2.1.2).

- Propose alternate routes and shipping schedules such that each shipment departs at a different time of the day if the licensee is planning to ship to the same destination for multiple shipping campaigns. Shipping operations should not appear routine, transparent, or predictable.

The licensee should submit the application for route approvals, including maps, tables, or charts in both paper copy and on a compact disc or other appropriate electronic storage media. The word processing file format used to develop the application should be the latest version of Microsoft Word or in a format readily convertible. In most cases, route approvals contain information designated as Safeguards Information (SGI). SGI must be stored, transmitted, and handled in accordance with the requirements of 10 CFR 73.21, "Protection of Safeguards Information: Performance Requirements," and 10 CFR 73.22, "Protection of Safeguards Information: Specific Requirements."

The information requested above should be submitted, in writing, along with the request for route approval to the following:

Director, Division of Security Policy
Office of Nuclear Security and Incident Response
U.S. Nuclear Regulatory Commission
Washington, DC 20555

Upon approval of a complete application, the NRC will approve road routes for 5 years and rail routes for 7 years of use. Port approvals will not carry a specific approval time; the port approval will be consistent with the mode by which the shipment enters or leaves the port. If a route is mixed (i.e., both rail and road), it will be approved for 5 years.[2] For renewals and amendments to existing routes, the licensee need only submit information containing changes from the previously approved route and reference the previous approval.

Before shipping by rail, the licensee should verify with the rail carrier that a route the NRC has previously approved is still the route posing the least overall safety and security risk. The licensee should submit a request to the NRC to amend the approved rail route if conditions warrant changes to certain segments of the route.

2.1.1. Route Selection Criteria

Licensees should apply the following criteria when selecting and developing primary and alternate routes for road, rail, and waterborne SNF shipments:

*   Minimized transit time. Routes should be selected to minimize the time or distance that the shipment is en route. When shipping by road, to the maximum extent practicable, shipments should be on primary highways with minimal use of secondary roads and should be made without intermediate stops except for refueling, rest, or an emergency. Routes that are neither minimum distance nor minimum time in transit are acceptable if the application documents that affected States and Tribes agree to the use of the route and that it meets DOT routing requirements or that DOT approval has been obtained.

*   Availability of swift local law enforcement agency (LLEA) response. Select routes that permit more timely LLEA response when assistance is requested.

*   Availability of safe haven locations. Routes should be selected that feature locations either on or near the approved route where, in the event of a security-related emergency, either security is present or from which the transport crew can notify or wait for the LLEA to respond. Section 2.1.2.2 discusses safe havens in greater detail.

*   Avoidance of tactically disadvantageous positions. Select routes that avoid, as much as practicable, passage through heavily populated areas and areas or terrain that would place the shipment or shipment escorts in significantly tactically disadvantageous positions (e.g., long tunnels or over bridges spanning heavily populated areas).

*   Availability of appropriate rest and refueling stops. Shipment routes should feature sufficient locations for rest and refueling stops to allow flexibility in adjusting schedules to accommodate unexpected situations.

*   Availability of good transportation safety design features. Road and rail routes featuring advanced safety design features (e.g., divided highways, guard rails, limited access highways for road shipments; high-grade track for rail shipments) are preferred over those with portions in disrepair or that are obsolete.

When selecting a route by road, the licensee should physically inspect the proposed route. Once a route is inspected, the licensee may use this information for future shipments over the

---

[2] NRC Regulatory Issue Summary 2006-01, "Expiration Date for NRC-Approved Spent Fuel Transportation Routes," dated January 24, 2006, sets the policy for 5-year approval of highway routes and 7-year approval of railway routes. (see Agencywide Documents Access and Management System (ADAMS) accession number ML052970029)

same roads. The licensee should determine whether a new inspection is warranted because of changing road conditions (e.g., significant road construction, demolition, or construction of rest areas). An inspection is also recommended when re-approval of an expiring route is being requested.

The road inspection team should consist of at least two personnel who travel the route to collect data and monitor the communications capability along the route. Two-person teams are recommended to perform these functions safely while driving. The road inspection should, at a minimum, include locating safe havens, fuel, and food stops for the carrier; developing LLEA contacts; and identifying potential driving problems along the route. When appropriate, the staff may conduct its own inspection of the proposed route or portions of it.

2.1.2. Format of Route Approval Applications

The licensee should submit a letter to the NRC requesting approval of the planned road or rail route(s) and U.S. port(s) where vessels carrying SNF are scheduled to stop. The information requested in Section 2.1 should be included in the licensee's request. The route overview and specific details should be attached to the letter and protected against unauthorized disclosure in accordance with 10 CFR 73.21 and 10 CFR 73.22. The following sections describe the route overview information and specific details expected in the submittal.

2.1.2.1. Route Plan and Mileage

The licensee should include the following information in the request to the NRC for approval of planned and alternative routes for road, rail, and waterborne shipments of SNF:

- origin and destination (i.e., specific locations and addresses)

- estimated travel time for each segment (i.e., estimated time for each segment under good conditions)

- location of planned or available stopover points for each route segment (e.g., for food, fuel, loading, or unloading)

- escort arrangements for each proposed route and alternative route

- firearms to be carried by armed escorts, if private guards are employed

- location and arrangements for staffing of movement control center

- plans for providing required communications capabilities

The licensee should include the following additional information in the request for approval of planned and alternative routes for road shipments:

- a mileage summary table that specifies the number of miles and accumulated distances, with driving directions, for each road by State (see the sample mileage summary table in Table A-1 of Appendix A)

- detailed route plan tables that specify the route segments by county and by participating Tribe's reservation for each State (see the sample detailed route plan table in Table A-2 of Appendix A)

The licensee should include a mileage summary by State in the request to the NRC for approval of planned and alternative routes for rail and waterborne shipments.

During the course of an actual shipment, circumstances may arise that preclude the use of some portion of the approved route for an extended period of time. In this case, detours may be taken provided that certain determinations are made and procedures are followed in lieu of having specific advance NRC approval of the revised portion of the route. These determinations and procedures are as follows:

- Licensee determines that a portion of the approved route is impassible or that conditions along the route will likely result in unplanned stops of 1 hour or more.

- Licensee determines that it is impractical to use an alternative route that the NRC has previously approved.

- Licensee determines that the detour selected will most likely allow for uninterrupted travel until the shipment is able to resume travel along the planned approved route or along a previously approved alternative route, or that the detour selected will most likely result in fewer stops or delays as compared to the planned approved route.

- Licensee ensures that contacts with the LLEA along the proposed detour route have been previously established or are established when a determination is made to use the detour route.

- Escorts notify the movement control center of the change in shipment itinerary, delays experienced along the planned or alternative route, and estimated duration of the detour.

- Licensee makes arrangements to provide any additional resources for physical protection needed to comply with the specific requirements of 10 CFR 73.37(c), (d), or (e).

- Licensee notifies the NRC Headquarters Operations Center about the detour.

Escorts should anticipate the need for making unplanned detours so that the shipment may keep moving without interruption. The shipment should not be stopped solely for the purpose of making new LLEA contacts along the detour or informing the movement control center that a detour is necessary, unless stopping is unavoidable, as may be the case if the escorts are unable to establish communications with either the LLEA or the movement control center. In these situations, the shipment should proceed to the nearest available safe haven before making the stop.

2.1.2.2. Safe Havens

Section 73.37(1)(vi) requires that the licensee include the locations of identified safe havens along road routes for temporary refuge or emergency assistance. These safe havens must be coordinated with the State(s) through which the shipment will pass. Safe havens do not apply to rail or waterborne shipments.

The licensee should identify safe havens along the route at routine intervals generally not exceeding 50 miles. If intervals greater than 50 miles are proposed, the licensee should provide justification for the greater distance. Also, safe havens should be as close to the highway as possible, easily accessible by the transportation vehicle, controlled, and well lit.

Examples of possible safe havens include truck stops, rest areas, military installations, highway patrol barracks, and weigh stations. Unless specifically approved by the State, hospitals, public libraries, and schools are not acceptable safe havens. The mile marker or exit number corresponding to each safe haven along the entire route in each State should be identified and incorporated into the detailed route plan tables.

2.1.2.3. Mapping Requirements

The licensee should provide a summary map that identifies the origination and destination facilities and the complete route that will be used for the shipment. In addition, the licensee should provide detailed individual State maps of the planned route and any alternate routes, indicating the locations of identified stops, boundary lines of any reservations of participating Tribes along the route, and the safe havens that have been coordinated with the State(s) through which the shipment will pass. Maps should be part of the word processing files or images easily integrated into the final approval document.

2.1.2.4. Contact Information

The route approval request should include the contact information for each LLEA jurisdiction along the planned route and, if applicable, the citizens band radio channels that each LLEA typically monitors. The licensee should provide the following:

- Twenty-four hour contact information available throughout the United States for the LLEAs responsible for every participating Tribe's reservation and county in each State along the route.

- Twenty-four hour, 10-digit emergency telephone number available throughout the United States for each State along the route.

- Contact information for the governor's designee and participating Tribal official's designee for advance notification of SNF shipments for the route segment in each State and participating Tribe's reservation along the route.

As part of the review, the staff will verify that the contact information is correct and current. Both the transport (i.e., driver and escorts) and the movement control center (see Section 2.4.1) will use the contact information. In addition to the telephone numbers listed above, an emergency number, such as 911, may be provided as a secondary number for the transport. However, the licensee should demonstrate in its application that its method of communication (e.g., cell phone, satellite phone) will reach the correct en-route 911 emergency response center. The licensee should demonstrate that an en-route 911 call will not be misdirected, thereby resulting in a delay of emergency response.

As stated earlier, routes will be approved for use over many years. However, during the approval period, State and participating Tribe contact information may change. Before each shipment of SNF, the licensee is responsible for ensuring that all the contact information along the route is current. Updating contact information is considered a change that does not reduce the effectiveness of the route and the licensee may make such updates to contact information without prior approval from the NRC.

The contact information specified above should be provided to drivers, escorts accompanying the shipment, and movement control center personnel monitoring the shipment. As required by Section 73.37(b)(4), the licensee must ensure that such personnel are instructed in the proper use of such information. If a large number of shipments are planned over the same route, spot checking LLEA telephone numbers and contacts during each trip would be sufficient for updating purposes. However, the licensee should perform more systematic verification procedures if shipments along the route are infrequent (e.g., more than 6 months apart). Route information verifications should be recorded on the shipment log.

The route applications should describe armed escort arrangements (including any transfers at State or local boundaries), contacts, and coordination with State governors or their designees. (While coordination with Tribal officials is not required, it is considered a best practice, as described more fully in Section 2.2, below.) The NRC may conduct surveys of proposed routes, including verifying the response capabilities of LLEAs along the routes.

2.1.2.5. *Communications Capability along the Route*

Poor communication may exist in areas along the shipment route. Therefore, the licensee should provide NRC with information about the primary and secondary communications capability along the route. The licensee should clearly identify areas along the route and any identified alternative route within a State or participating Tribe's reservation where poor communications may exist. This information should be incorporated into the detailed route plan tables.

2.2.   **Preplanning and Coordination of Shipments**

In addition to meeting the requirements in 10 CFR 73.37(b)(1)(vi) to obtain advance approval of the proposed route, the licensee should preplan and coordinate other aspects of the shipment before it begins.

Section 73.37(b)(1) requires the licensee to preplan and coordinate shipment details no later than two weeks before the SNF shipment or before the first shipment of a series of shipments, with the governor, or the governor's designee, of each State through which or across whose boundaries the shipment will pass. The objectives of this activity are to:

*   Ensure minimal intermediate stops and delays.

*   Discuss the State's intention to provide law enforcement escorts.

*   Arrange for sharing positional information about a shipment when a State requests it.

*   Develop primary and, where applicable, alternate route information, including the identification of safe havens.

*   Arrange the response to an emergency or call for assistance along the shipment route with the LLEA.

The licensee also shall coordinate shipment schedules and itineraries with the receiver at the final delivery point to ensure that the receiver is present to accept the shipment. Additionally, if applicable, any transfer of custody (in which armed escorts transfer responsibility for the shipment) during the shipment must be preplanned and coordinated to ensure that the required written certification of the transfer of custody is properly documented and completed.

The licensee is also required to protect SGI in accordance with 10 CFR 73.21 and 10 CFR 73.22. Preplanning and coordination activities should address how this information will be protected, especially the determination of the trustworthiness and reliability of any individual whose assigned duties will provide access to the SNF shipment information.

The licensee should document the preplanning and coordination activities and protect sensitive information as SGI as specified in 10 CFR 73.21 and 10 CFR 73.22. The types of activities that should be documented include, but are not limited, to timelines for outreach to States (e.g., meetings, teleconferences), summaries of planning meeting discussions, and lists of personnel contacted.

Although not a requirement, a best practice also would be to preplan and coordinate shipment details no later than 2 weeks before the shipment or before the first shipment of a series of shipments with the Tribal official, or the Tribal official's designee of each participating Tribe's reservation through which or across whose boundaries the shipment will pass.

2.2.1. Avoidance of Intermediate Stops

To satisfy 10 CFR 73.37(b)(1)(iv)(A), the licensee should ensure that intermediate stops and delays are kept to a minimum. Intermediate stops should be avoided to the extent practicable because it is recognized that a shipment is generally more vulnerable to attempts at theft, diversion, or radiological sabotage when stationary. Scheduled stops for SNF shipments should be justified by a security or operational need for making the stop. When possible, stops should be planned to serve multiple purposes and to avoid set patterns in the times and places they occur.

2.2.2. Procedures at Stops

Shipments are vulnerable to attack while stationary. In accordance with 10 CFR 73.37(b)(3)(vii)(C), the licensee shall ensure that at least one armed escort other than the driver maintains constant visual surveillance of the shipment and reports to the movement control center at regular, pre-set intervals not to exceed 30 minutes during periods when the shipment vehicle is stopped or the shipment vessel is docked. Maintaining surveillance of the shipment during these periods is intended to ensure that any attack would be detected as early as possible so that the adversary would have only a minimal amount of time to attempt theft, diversion, or sabotage of the SNF before response forces arrive.

2.2.2.1. Stops during Road Shipments

The licensee should ensure that, when a shipment has stopped, the armed escorts maintain visual observation of the transport vehicle to detect tampering with or unauthorized access to the shipment. The visual observation by the armed escorts should include periodic walks around the vehicle. Multiple armed escorts should be separated to prevent being simultaneously incapacitated by a single adversary act. Additionally, 10 CFR 73.37(c)(2) requires, as permitted by law, that each armed escort be equipped with a minimum of two weapons (e.g., a handgun and a rifle or shotgun). This requirement does not apply to LLEA personnel performing escort duties.

The licensee should ensure that the driver and armed escorts assigned to the vehicle surveillance are provided with the following:

- the means to immobilize the transport vehicle

- the means to communicate with the movement control center and LLEA

- the route overview information containing specific, direct LLEA contacts, other than dialing 911

If the LLEA is not directly involved in the escort, the LLEA should be advised of the presence of the shipment within its jurisdiction and the location, purpose and projected duration of the stop. The LLEA should be requested to be alert for possible assistance calls during the stop.

Whether occupied or not, the doors and windows of the transport vehicle should be closed and locked to impede unauthorized access. Before making the stop, the escorts should ensure that communications equipment is properly operating and that LLEA contact information is readily available.

All of the armed escorts should participate in the surveillance, as well as any drivers or accompanying individuals not engaged in essential activities associated with the purpose of the stop.

2.2.2.2. Stops during Rail Shipments

Procedures for maintaining visual surveillance of a rail shipment reflect the same concerns that form the basis of surveillance procedures for road shipments.

During all stops, the licensee should ensure that the rail car is placed under surveillance by two armed escorts positioned on the train within view of the shipment car. If the armed escorts cannot or do not remain on the train, they should relocate to a protected position with an unobstructed view of the shipment car and possible approaches to it. The escorts should have immediate access to communications equipment and the route overview information containing LLEA contacts and should be alert to external activities that may indicate a possible threat to the safety or security of the shipment.

When practical, multiple armed escorts should be separated to prevent being simultaneously incapacitated by a single adversary act. Additionally, 10 CFR 73.37(d)(2) requires, as permitted by law, that each armed escort be equipped with a minimum of two weapons (e.g., a handgun and a rifle or shotgun). This requirement does not apply to LLEA personnel performing escort duties.

2.2.2.3. Stops during Waterborne Shipments

When incoming and outgoing waterborne shipments of SNF are temporarily located on a dock or aboard a docked vessel, 10 CFR 73.37(e) requires the licensee to ensure surveillance by either of the following methods:

- two armed escorts stationed either on board the shipment vessel or on the dock at a location that will permit observation of the shipment vessel or the points of access to the shipment, if it is located in an enclosed cargo compartment

- a member of an LLEA, equipped with normal radio communications, stationed either on the vessel or on the dock

Section 73.37(b)(3) requires surveillance to be provided continuously for outgoing shipments from the time the shipment arrives at the terminal aboard a road or rail transport vehicle to the time the loaded shipment vessel leaves the dock. Incoming shipments must be placed under continuous surveillance from the time the shipment vessel docks to the time it is loaded onto a road or rail transport vehicle and begins movement from the terminal, at which time the physical protection requirements governing road or rail transport shall apply.

These procedures also apply to intermediate stops at U.S. ports during which the shipment remains on the shipment vessel. Visual surveillance of the shipment should be accomplished by direct observation of the shipment or by observing points of access to the shipment if it is located in an enclosed cargo compartment.

Armed escorts performing the surveillance should have immediate access to communications equipment and the route overview information containing LLEA contacts and should be alert to external activities that may indicate possible threats to the safety or security of the shipment. Additionally, 10 CFR 73.37(e)(2) requires as permitted by law that each armed escort be equipped with a minimum of two weapons (e.g., a handgun and a rifle or shotgun). This requirement does not apply to LLEA personnel performing escort duties.

2.3. Advance Notification of Shipments

Both 10 CFR 73.37, "Requirements for Physical Protection of Irradiated Reactor Fuel In Transit," and 10 CFR 73.72, "Requirements for Advance Notice of Shipment of Formula Quantities of Strategic Special Nuclear Material, Special Nuclear Material of Moderate Strategic Significance, or Irradiated Reactor Fuel" contain specific requirements for advance notification of SNF shipments. Before transporting SNF outside the confines of the licensee's facility or other place of use or storage, the licensee shall provide notification to the NRC. The licensee shall also provide notification to the governors of the States, or the governors' designees, and after June 11, 2013, to the Tribal officials of participating Tribes or the Tribal officials' designees of any SNF shipment moving through or across the boundary of the applicable States or a participating Tribe's reservation. A licensee that conducts an onsite transfer of SNF that does not travel on or cross a public highway is exempt from the requirements for advance notifications for that transfer.

2.3.1. Advance Notification of the State(s) and Tribes

Section 73.37(b)(2) requires that advance notification delivered to a State or participating Tribe by mail be postmarked at least 10 days before transport of a shipment within or through the State or the participating Tribe's reservation. If a notification is delivered by any other method, it must reach the office of the governor, or the governor's designee, and the Tribal official or the Tribal official's designee at least 7 days before the scheduled transport of a shipment within or through the State or the participating Tribe's reservation.

Section 10 CFR 73.37(b)(2) requires that advance notification be in writing and include the following information:

- name, address, and telephone number of the shipper, carrier, and receiver of the shipment and the license number of the shipper and receiver

- a description of the shipment as specified by DOT in 49 CFR 172.202, "Description of Hazardous Material on Shipping Papers," and 49 CFR 172.203(d)

- a list of the routes to be used within the State or participating Tribe's reservation

A best practice would be to provide within the advance notification a unique identifier for each SNF shipment. An example of an identifier is "ANO-1201," in which "ANO" would indicate the shipper (in this example, Arkansas Nuclear 1) and "1201" would indicate the first SNF shipment from ANO in 2012. Referring to this unique shipment identifier—instead of resorting to more obvious identifiable information when discussing the shipment in subsequent communications with the NRC, States, and participating Tribes—would help prevent inadvertent passing of SGI information through an unsecure method.

Section 73.37(b)(2)(iii) requires that the following information be contained in a separate enclosure to the written notification:

- the estimated date and time of departure from the point of origin of the shipment

- the estimated date and time of entry into the State or the participating Tribe's reservation

- the estimated date and time of arrival of the shipment at the destination

- a statement that schedule information must be protected in accordance with the provisions of 10 CFR 73.21 and 10 CFR 73.22 until at least 10 days after the shipment has entered or originated within the State or participating Tribe's reservation, for the case of a single shipment whose schedule is not related to the schedule of any subsequent shipment

- a statement that schedule information must be protected in accordance with the provisions of 10 CFR 73.21 and 10 CFR 73.22 until 10 days after the last shipment in the series has entered or originated within the State or participating Tribe's reservation and an estimate of the date on which the last shipment in the series will enter or originate within the State or participating Tribe's reservation, for the case of a shipment in a series of shipments whose schedules are related.

In accordance with 10 CFR 73.37(b)(2)(vi), the licensee must retain a copy of the advance notification(s), any revision notice(s), and cancellation notice(s) as a record for 3 years.

The State and participating Tribe contact information, including telephone number and mailing addresses of governors, governors' designees, Tribal officials, and Tribal officials' designees is available on the NRC Web site at http://nrc-stp.ornl.gov/special/designee.pdf.

A list of the mailing addresses of governors, governors' designees, Tribal officials, and Tribal officials' designees is also available upon request from the following:

Director, Division of Intergovernmental Liaison and Rulemaking
Office of Federal and State Materials and Environmental
  Management Programs
U.S. Nuclear Regulatory Commission
Washington, DC 20555-0001

2.3.2. Advance Notification of the NRC

A licensee that transports, ships, imports, or exports SNF must make advance notifications to the NRC at various times before and after the shipment begins, as specified in 10 CFR 73.72.

2.3.2.1. 10-Day Advance Notice

In accordance with 10 CFR 73.72, the licensee must notify the NRC at least 10 days before transport of the shipment begins at the shipping facility. The licensee shall provide written notification to the NRC Director, Division of Security Policy, Office of Nuclear Security and Incident Response. Notifications that contain SGI must be made in compliance with the requirements in 10 CFR 73.21 and 10 CFR 73.22.

The notification should include the following information:

- the names, addresses, and telephone numbers of the shipper, receiver, and carrier(s)

- the physical form of the SNF, quantity, type of reactor, and original enrichment

- a list of the modes of shipment, transfer points, and routes to be used

- the estimated time and date that the shipment will begin

- for exports and imports of SNF: the estimated time and date that each country along the route is scheduled to be entered

- the estimated time and date of arrival of the shipment at the destination

2.3.2.2. 2-Day Advance Notice, 2-Hour Notice, and Delivery Receipt Confirmation

Section 73.72(a)(4)requires the licensee to make additional notifications after the 10-day advance notification. The NRC Headquarters Operations Center shall be notified about the shipment status by telephone at the phone numbers listed in Appendix A, "U.S. Nuclear Regulatory Commission Offices and Classified Mailing Addresses," to 10 CFR Part 73, "Physical Protection of Plants and Materials." SGI notifications shall be made by secure telephone. The notifications shall take place at the following intervals:

- at least 2 days before the shipment begins
- 2 hours before the shipment begins
- once the shipment is received at its destination

2.3.3. Notification of Schedule Deviations

Section 73.37(b)(2)(iv) requires the licensee to provide a revised notice if the SNF shipment schedule changes by more than 6 hours from the notification provided to the State and participating Tribes. The licensee must notify, by telephone or other means, a responsible individual in the office of the governor, or in the office of the governor's designee, and in the office of the Tribal official, or in the office of the Tribal official's designee, about the schedule change and shall inform that individual of the number of hours of advance or delay relative to the written schedule previously furnished. In addition, the licensee must notify the NRC Headquarters Operations Center by telephone at the phone numbers listed in Appendix A to 10 CFR Part 73. SGI notifications must be made by secure telephone.

If a shipment is cancelled after the advance notifications have been made, 10 CFR 73.37(b)(2)(v) requires that the licensee send a cancellation notice to the governor of each State, or to the governor's designee, and to each Tribal official or to the Tribal official's designee previously notified. The cancellation notice must also be sent to the NRC's Director, Division of Security Policy, Office of Nuclear Security and Incident Response. To avoid situations in which State and Tribal resources are committed unnecessarily, the notice should be transmitted as soon as possible after the decision has been made to cancel the shipment. The licensee shall state in the notice that it is a cancellation and identify the advance notification being canceled. While a telephone call is the preferred method of communicating a notice of cancellation, the licensee must ensure that the notice does not contain SGI. As required in 10 CFR 73.22, an NRC-approved secure electronic device, such as a facsimile or secure telephone must be used if the notice contains SGI (e.g., detailed information related to rescheduling the shipment).

Section 73.37(f) requires the licensee that made the arrangements for the SNF shipment to immediately conduct an investigation, in coordination with the receiving licensee, when a shipment is lost or unaccounted for after the designated no-later-than arrival time in the advance notification (i.e., the estimated arrival time). While the investigation is ongoing, all information about the investigation shall be protected against unauthorized disclosure as required by10 CFR 73.21 and 10 CFR 73.22. Once the investigation is complete, the records shall be designated and managed in accordance with these regulations.

2.4. Communications

2.4.1. Movement Control Centers

Section 73.37(b)(3) requires the licensee to establish a movement control center that is continuously staffed (24 hours a day, 7 days a week) by at least one individual with the authority to coordinate physical protection activities during the active shipment. This individual(s) is responsible for monitoring the progress of the SNF shipment to ensure its continued integrity and to notify the appropriate agencies in accordance with 10 CFR 73.71, "Reporting of Safeguards Events," should such an event arise. As required by Section 73.37(b)(3)(v)(B), the licensee shall develop, maintain, revise, and implement written procedures that document the duties of the movement control center personnel.

The movement control center operator should maintain a record of the status reports received for inclusion in the shipment log (see Section 2.7 of this document). The operator also should maintain a record of any deviation from the planned shipment itinerary, any significant incidents that occur, and reports made to the NRC and others. The operator should have a copy readily available of the route data for the shipment and should request assistance from the LLEA in whose jurisdiction the shipment is believed to be located if the transport does not report in on schedule or cannot be contacted, or if a request for assistance is received from the shipment escorts. (Escort procedures should require escorts to request LLEA assistance directly in cases when it is operationally feasible to do so.) The movement control center should be able to directly contact appropriate LLEA in each jurisdiction; the center should not rely on calls to 911 systems to provide aid to shipments of SNF.

A procedure or device should be used to ensure that the movement control center operator becomes aware immediately when a report from the shipment escorts is overdue. The operator can carry out other work while a shipment is en route, provided that the other work does not interfere with prompt response to incoming messages from the transport or with other security-related duties.

2.4.2. Periodic Contacts with Movement Control Center

Section 73.37(b)(3)(vii)(B)requires escorts for road and rail shipments to notify the movement control center at random intervals not to exceed 2 hours while in motion. Section 73.37(b)(3)(vii)(C)requires at least one armed escort to remain alert, maintain visual surveillance of the shipment, and periodically report to the movement control center at regular intervals not to exceed 30 minutes during periods when the shipment vehicle is stopped or the shipment vessel is docked. Notification calls also should be made when the shipment undergoes a significant change in status or when delays are encountered that will result in changes in the shipment schedule or itinerary (e.g., when making an unscheduled stop or detour, or resuming travel).

Section 73.37(b)(3)(v)requires the licensee to develop, maintain, revise, and implement written communication protocols for use during an SNF shipment. These protocols should include a strategy for the use of authentication and duress codes, the management of refueling or other stops and detours, and the loss of communications, temporary or otherwise.

2.5.   Arrangements with Local Law Enforcement Agencies

In 10 CFR 73.37(b)(1)(v), the NRC requires that licensees make arrangements with LLEA for their response to a security-related emergency or a call for assistance along the routes of road and rail shipments, and at U.S. ports where vessels carrying SNF shipments are docked. This requirement is designed to provide for rapid LLEA assistance in the event of a security-related emergency or a call for assistance. It is also intended to ensure that the selected LLEAs along the route are familiar with the types of situations to expect when responding to such calls. Section 73.37(b)(4)(iv)requires the licensee to notify LLEAs of suspicious activities and to request assistance without delay, but no later than 15 minutes after discovery of the threat.

The licensee's application for route approval should describe armed escort arrangements, contacts, State governor coordination, and, if applicable, points where armed escorts transfer responsibility for the shipment. The NRC may conduct surveys of proposed routes, including verifying the response capabilities of LLEAs along the routes.

2.6.   Armed Escorts

Section 73.37(b)(3)(v)(B)requires that the licensee develop, maintain, revise, and implement written transportation physical protection procedures that address the duties of the armed escorts.

The licensee must arrange for armed escorts in accordance with 10 CFR 73.37. Depending on State requirements or availability, each State may provide the armed escorts, or the applicant may need to contract armed escorts. This requirement does not pre-empt existing Federal or State statutes or regulations on the use of weapons and use of deadly force.

As required by Section 73.37(b)(1)(i), each licensee shall ensure that each armed escort is instructed in the use of force sufficient to counter the force directed at that individual, including the use of deadly force when the armed escort has a reasonable belief that it is necessary in self-defense or in the defense of others, or any other circumstances, as authorized by applicable Federal and State laws. This requirement does not apply to members of LLEAs performing escort duties.

The licensee should evaluate the State requirements for armed escorts moving firearms from State to State in accordance with 10 CFR 73.37(b)(1). All armed escorts, with the exception of LLEA personnel, must comply with and be trained in the appropriate State rules governing the use of firearms.

In addition, as permitted by law, each armed escort shall be equipped with a minimum of two weapons. The NRC recommends that each weapon provide separate and distinct response capabilities (e.g., a handgun and a rifle or shotgun). This requirement does not apply to LLEA personnel performing escort duties.

Section 73.37(b)(3)(vii), requires that armed escorts for road, rail, or waterborne shipments of SNF successfully complete the training required by Appendix D, "Physical Protection of Irradiated Reactor Fuel in Transit, Training Program Subject Schedule," to 10 CFR Part 73, including the equivalent of the weapons training and qualifications program required of guards, as described in Sections III and IV of Appendix B, "General Criteria for Security Personnel," to Part 73. These training and qualification requirements do not apply to LLEA personnel or ship's officers serving as unarmed escorts. Appendix D contains a list of subjects the course should cover. The detailed material taught under each subject heading should be adapted to the particular modes of transportation used for the shipment that the armed escort will be assigned to protect. The licensee may adjust the length of the course depending on the background and the experience of the individuals selected to be armed escorts. The licensee should be prepared to demonstrate the efficacy of the training program to comply with the NRC inspection and enforcement program. One way to accomplish this is to maintain records of the licensee's evaluations (written tests, field tests, or observation of performance) of the trainees' progress in the instruction program and any remedial training required.

Training for the armed escorts also should address the need for the escorts to anticipate making unplanned detours so that the shipment may keep moving without interruption. The shipment should not be stopped solely for the purpose of making new LLEA contacts or communicating with the movement control center. In some situations, however, stops may be unavoidable. In these cases, the shipment should proceed to the nearest available safe haven before making the stop.

LLEA personnel serving as escorts are considered to be adequately trained to carry out the escort duties they are expected to perform and are not required to undergo the training specifically required for private armed escorts. However, a member of an LLEA should be briefed on the shipment procedures as necessary to perform the escort functions.

Section 73.37(b)(3)(vii) requires shipment escorts to provide notifications to the movement control center at random intervals, not to exceed 2 hours, to advise of the status of road and rail shipments while in motion, and at regular intervals not to exceed 30 minutes when the shipment vehicles are stopped or the shipment vessels are docked at U.S. ports. These requirements are covered in Section 2.4.

2.7. Shipment Log

Section 73.37(b)(3)(iv)requires the movement control center personnel and the armed escorts to maintain a written log for every SNF shipment. The log should contain information describing the shipment and any significant events that occur during the shipment. Entries in the log should be coordinated between the armed escorts and the movement control personnel monitoring the shipment. The purpose of this requirement is to ensure that protection requirements have been carried out properly, to allow NRC staff to identify areas where protection improvement is needed, and to ensure the availability of an accurate record of the shipment to assist in resolving any questions of public concern about the shipment that could arise after its completion.

The shipment log should include the following types of information:

*   names of shipping and receiving organizations, carriers, escorts, drivers (for road shipments), chief engineers (for rail shipments), and vessels and their masters (for waterborne shipments)

*   origin, destination, and approved route (copy of route overview)

*   general description of cargo and shipping containers

*   dates and times of departure and arrival (planned and actual)

*   dates, times, and locations of stopovers and custody transfers

*   identification of the movement control center and its staff

*   dates, times, and locations of status reports made by shipment escorts to the movement control center, and schedule of expected status reports, as well as the name of the escort making the report and the movement control operator taking the report

*   deviations from the planned route

*   other abnormal occurrences with regard to route, equipment, vehicles, personnel, weather, traffic, or threats

In accordance with 10 CFR 73.37(b)(3)(iv), the licensee shall retain the shipment log for 3 years following completion of the shipment and make the log available to the NRC upon request.

2.8. Procedures, Training, and Control of Information

To comply with 10 CFR 73.37(b)(3)(v), the licensee shall develop, maintain, revise, and implement written procedures for transportation physical protection to address the following issues:

*   access controls to ensure that no unauthorized persons have access to the shipment and SGI

- duties of the movement control center personnel, drivers, armed escorts, and other individuals responsible for the security of the shipment

- reporting of safeguards events in accordance with 10 CFR 73.71

- communication protocols that include a strategy for the use of authentication and duress codes; the management of refueling or other stops, detours, and the loss of communications, temporary or otherwise; and communication with the NRC in accordance with 10 CFR 73.4, "Communications"

- normal-condition operating procedures

To comply with 10 CFR 73.37(b)(3)(vi), each licensee should develop and implement a method or system to ensure that records of transportation physical protection procedures are maintained for at least 3 years after the close of the period for which the licensee possesses the SNF. If any portion of the procedure is superseded, the licensee should ensure the retention of the superseded material for 3 years after each change.

2.8.1. Procedures for Coping with Threats and Emergencies

Section 73.37(b)(4)(i)requires the licensee to establish, maintain, and follow written contingency and response procedures to deal with threats, theft, and radiological sabotage related to SNF in transit.

The contingency plan and procedures shall be retained as a record for 3 years after the close of the period for which the licensee possesses the SNF in accordance with 10 CFR 73.37(b)(4)(iii). Additionally, each licensee shall ensure that information related to the physical protection procedures for coping with threats and emergencies is protected in accordance with the requirements in 10 CFR 73.21 and 10 CFR 73.22.

Upon detection of the abnormal presence of unauthorized individuals, vehicles, or vessels in the vicinity of an SNF shipment or upon detection of a deliberately induced situation that has the potential for damaging an SNF shipment, the contingency plan and procedures should direct the armed escort to do the following:

- Determine whether a threat exists.

- Assess the extent of the threat, if any.

- Implement the approved contingency and response procedures developed in accordance with these requirements.

- Make the necessary tactical moves to prevent or impede theft, diversion, or radiological sabotage of SNF.

- Inform the LLEA of the threat and request assistance without delay, not exceeding 15 minutes after discovery.

It is important to note that armed escorts are neither required nor expected to take offensive action against aggressors (e.g., actively pursuing and apprehending suspected aggressors), but they are expected to assume a defensive posture to delay and impede attempts at theft, diversion, and radiological sabotage of SNF shipments as appropriate, considering threat characteristics, shipment characteristics, and the primary requirement for personnel to provide for their own safety. It is imperative for armed escorts, drivers, or other accompanying personnel to contact response personnel without delay as soon as they detect a threat to the shipment or to themselves.

If a shipment is lost or unaccounted for after the designated no-later-than arrival time in the advance notification, the licensee that made the arrangements for the shipment of SNF shall immediately conduct an investigation as required by 10 CFR 73.37(f). The receiving licensee shall also assist in the investigation. While the investigation is ongoing, all information about the investigation shall be protected against unauthorized disclosure as specified in 10 CFR 73.21 and 10 CFR 73.22. Once the investigation is complete, the records shall be designated and managed in accordance with 10 CFR 73.21 and 10 CFR 73.22.

2.8.2. Training of Personnel Associated with Shipments

To ensure that personnel associated with SNF shipments are adequately prepared for all possible scenarios, the licensee should develop and implement a documented training program that covers the procedures set forth in 10 CFR 73.37(b)(3)(v) and 10 CFR 73.37(b)(4), to include transportation physical protection procedures, normal operations procedures, and contingency and response procedures. See Section 2.6 above for information on specific training requirements for armed escorts.

2.8.3. Control of Shipment Information

The procedures for transportation physical protection shall be protected against unauthorized disclosure as specified in 10 CFR 73.21 and 10 CFR 73.22. The licensee should ensure that procedures are developed and implemented to protect shipment information, procedures for transportation physical protection, and all required records against unauthorized disclosure. As required by 10 CFR 73.38, "Personnel Access Authorization Requirements for Irradiated Reactor Fuel in Transit," each licensee must also determine the trustworthiness and reliability of any individual whose assigned duties provide access to SNF shipment information.

3.   SHIPMENTS BY ROAD

This chapter provides guidance on the U.S. Nuclear Regulatory Commission (NRC) requirements specific to physical protection of spent nuclear fuel (SNF) shipments made by road. Requirements discussed in this chapter are in addition to and not a substitute for requirements listed elsewhere in this NUREG that are applicable to physical protection of all SNF shipments regardless of the mode of transport (road, rail or water). The requirements for road shipments apply to all SNF shipment vehicles in a convoy, unless noted otherwise.

Information on 10 CFR 73.37(c)(3), (4), and (6) shall be protected against unauthorized disclosure as specified in 10 CFR 73.21, "Protection of Safeguards Information: Performance Requirements," and 10 CFR 73.22, "Protection of Safeguards Information: Specific Requirements."

3.1.   Protection of Road Shipments

In addition to the general requirements in 10 CFR 73.37(b), the physical protection system for any portion of an SNF shipment made by road must meet the requirements in 10 CFR 73.37(c).

When SNF is transported by road, the transport vehicle must be accompanied by at least two armed escorts. This requirement can be met in one of two ways:

- The transport vehicle is occupied by at least two individuals, one of whom serves as an armed escort, and the transport vehicle is escorted by an armed member of the local law enforcement agency (LLEA) in a mobile unit of this agency.

- The transport vehicle is led and trailed by escort vehicles containing at least one armed escort each, for a minimum of three vehicles in the convoy.

Section 73.37(c)(2) requires, as permitted by law, that each armed escort carry a minimum of two weapons. The NRC recommends that each weapon provide separate and distinct response capabilities (e.g., a handgun and a rifle or shotgun). This requirement does not apply to LLEA personnel performing escort duties.

3.2.   Communications for Road Shipments

The purposes of communication requirements for SNF shipments by road are to: (1) provide the escorts with the capability to call for assistance when necessary, either directly to the LLEA or indirectly through the movement control center, (2) permit personnel at the movement control center to monitor the progress of the shipment, (3) provide the escorts with a way to quickly develop new LLEA contacts and obtain new route information when unexpected detours become necessary, and (4) provide a way to coordinate the movement of transport and escort vehicles when more than one vehicle is used in the shipment.

In 10 CFR 73.37(c)(3), the NRC requires the transport vehicle and all escort vehicles to be equipped with redundant communications abilities that provide for two-way communications among the transport vehicle, escort vehicles, movement control center, LLEA, and each other at all times. Alternate communication methods should not be subject to the same failure modes as the primary communication method. The transport vehicle and armed escorts must use a method capable of contacting the emergency phone numbers provided in the approved route.

If a citizens' band (CB) radio is being used, all escorts and drivers should be knowledgeable of the specific CB channels normally monitored by the LLEA and Radio Emergency Associated Communications Teams (REACT) as they pass through different jurisdictions on the shipment route. If an emergency arises, the occupants of each transport or escort vehicle are then capable of independently contacting the LLEA by CB radio. This information should be indicated on the route overview.

In addition to voice communication, 10 CFR 73.37(c)(6) requires licensees to ensure that shipments are continuously and actively monitored by a telemetric position monitoring system or an alternative tracking system reporting to a movement control center. The movement control center shall provide positive confirmation of the location, status, and control over the shipment. Additionally, the movement control center shall implement preplanned procedures in response to deviations from the authorized route or a notification of actual, attempted, or suspicious activities related to the theft, loss, or diversion of a shipment. These procedures will include, but are not limited to, the identification of and contact information for the appropriate LLEA along the shipment route.

3.3.   Immobilization of Transport Vehicle

As required by 10 CFR 73.37(c)(4), the transport vehicle must be equipped with NRC-approved features that permit immobilization of the cab or cargo-carrying portion of the vehicle. This requirement applies equally to all transport vehicles used in a single shipment. In this requirement, immobilization means rendering the loaded transport vehicle incapable of movement under its own power.

The purpose of this requirement is to deny an adversary who may succeed in gaining control of a transport vehicle the ability to move or flee with the vehicle. The immobilization technique should be implemented only when it is apparent that an attempt is being made to gain unauthorized control over the shipment. Immobilization should not be initiated in a way that would endanger the driver, escorts, or members of the public.

Immobilization procedures should be included in the contingency and response procedures developed in accordance with 10 CFR 73.37(b)(4). As required by 10 CFR 73.37(c)(5), operation of the immobilization technique and the procedures governing its use must be covered in both the training course for escorts and the familiarization program for drivers.

3.3.1. Immobilization Device Performance Criteria

The immobilization device should meet each of the following criteria:

*   The immobilization device and procedure should be able to be operated and performed from inside the cab of the transport vehicle by one person.

*   Immobilization should occur shortly (several seconds) after the immobilization procedures are initiated.

*   After immobilization is accomplished, skilled technical personnel should require at least one-half hour to return the transport vehicle to normal operating conditions. It should not be possible, by coercion of the drivers or escorts, for an adversary to bypass the effects

of the immobilization or to significantly shorten the time needed to make the transport operational again.

- The device should pose no significant safety hazard before, during, or after the immobilization.

3.3.2. Immobilization Device Design

Devices employed to effect immobilization may be mechanical or electrical. They should be relatively simple and reliable to operate, so that they can be activated quickly under stressful conditions. The following are some techniques that may form the basis for acceptable immobilization procedures:

- severing the main wire harness under the dashboard

- disabling a critical portion of the ignition system by overloading or dismantling a key component of the ignition system or starting system

- disabling the gear shifting mechanism

- using an electronic ignition control system with a procedurally irreversible time delay feature

3.3.3. Format of Approval Request

The licensee shall submit a letter to the NRC requesting approval of the intended immobilization device in advance. The device specification shall be attached to the letter and protected in accordance with 10 CFR 73.21 and 10 CFR 73.22.

3.4. Training Program for Drivers

Transport vehicle drivers are required in 10 CFR 73.37(c)(5) to be familiar with and capable of implementing transport vehicle immobilization, communications, and other security procedures.

The extent to which drivers may become involved in the physical protection of the shipment depends on the arrangements made between the carrier and the shipper or receiver (licensee). The greatest degree of driver involvement would occur when the driver is also a fully trained armed escort and alternately assumes driving and physical protection responsibilities with other armed escorts. However, in other circumstances, the driver may have only minimal responsibilities in protecting the shipment. In all cases, as required by Section 73.37(c)(5), the licensee must ensure that the driver is familiar with the basic security functions of transport vehicle immobilization, redundant communications systems, and any other security procedures that would affect the driver's operation of the transport vehicle.

The licensee should review the roles and responsibilities of the driver and ensure that the required training is conducted before the SNF shipment commences.

4. SHIPMENTS BY RAIL

This chapter provides guidance on the U.S. Nuclear Regulatory Commission (NRC) requirements specific to physical protection of spent nuclear fuel (SNF) shipments made by rail. Requirements discussed in this chapter are in addition to and not a substitute for requirements listed elsewhere in this NUREG that are applicable to physical protection of all SNF shipments regardless of the mode of transport (road, rail or water).

The transport vehicle in the case of a rail shipment is the shipment car carrying one or more SNF packages. The requirements for rail shipments should be understood to apply to the shipment cars as a group, unless noted otherwise.

Information identified in 10 CFR 73.37(d)(3) and 10 CFR 73.37(d)(4) shall be protected against unauthorized disclosure as specified in 10 CFR 73.21, "Protection of Safeguards Information: Performance Requirements," and 10 CFR 73.22, "Protection of Safeguards Information: Specific Requirements."

4.1. Protection of Rail Shipments

In addition to the general requirements of 10 CFR 73.37(b), 10 CFR 73.37(d) requires the physical protection system for any portion of an SNF shipment transported by rail to provide the following:

- A shipment car is accompanied by two armed escorts (who may be members of a local law enforcement agency (LLEA)), at least one of whom is stationed at a location on the train that will permit observation of the shipment car while in motion.

- As permitted by law, each armed escort shall be equipped with a minimum of two weapons. The NRC recommends that each weapon provide separate and distinct response capabilities (e.g., a handgun and a rifle or shotgun). This requirement does not apply to LLEA personnel performing escort duties.

The armed escorts may be private guards or members of the LLEA. The armed escorts shall be trained in accordance with Appendix D, "Physical Protection of Irradiated Reactor Fuel in Transit, Training Program Subject Schedule," to 10 CFR Part 73, "Physical Protection of Plants and Materials," and should be thoroughly familiar with all security requirements. The armed escorts should be alert to recognize any situations that might constitute a threat to the safety or security of the shipment. See Section 2.6 above for more information on specific training requirements for armed escorts.

A copy of the route overview data (i.e., route identification, mileage data, LLEA identification, jurisdiction, and response centers, LLEA telephone numbers, communication channels monitored by LLEAs, cellular and satellite phone coverage along the route) should be readily available to the escorts at all times.

The escorts should maintain close cooperation with the train's crew to ensure adherence as close as practicable to the shipment schedule and to ensure that the crew remains aware of all security requirements as the shipment progresses.

4.2. Communications for Rail Shipments

The purposes of the communication requirements for SNF shipments by rail are to: (1) provide escorts with the capability to call for assistance when necessary, (2) permit personnel at the movement control center to monitor the progress of the shipment, and (3) provide the escorts with a way to develop new LLEA contacts quickly and obtain new route information when unexpected detours become necessary.

Section 73.37(d)(3) requires the train's operator(s) and each escort to be equipped with redundant communications abilities that provide for two-way communications among the train operator(s), escorts, movement control center, LLEAs, and one another at all times. Alternate communication methods should not be subject to the same failure modes as the primary communication method. The use of telephones in call boxes located along the tracks generally is not acceptable as one of the communication methods because the telephones may not be available if the train is forced to make an emergency stop. However, call boxes could be relied on for short intervals during which cellular and satellite phone service is unavailable.

If the train's communications system is used, complementary communication capabilities should be provided to ensure that the escorts will have immediate access to the communications equipment when necessary, if it is located a considerable distance from the escort.

The shipment route overview should indicate areas where cellular, satellite, or other equivalent communications equipment may not be effective along the route. For such areas, alternate means of communication should be planned.

All equipment used to satisfy the communications requirements in 10 CFR 73.37(d)(3) should be maintained properly and checked for proper operation before the shipment begins to ensure that everything is in good operating condition before commencing the shipment.

As required by 10 CFR 73.37(d)(4), the licensee must ensure that rail shipments are monitored by a telemetric position monitoring system or an alternative tracking system reporting to the licensee, third-party, or railroad movement control center. The movement control center shall provide positive confirmation of the location of the shipment and its status. The movement control center shall implement preplanned procedures in response to deviations from the authorized route or to a notification of actual, attempted, or suspicious activities related to the theft or diversion of a shipment. These procedures will include, but are not limited to, the identification of and contact information for the appropriate LLEA along the shipment route.

The operator in the movement control center is required to monitor the shipping container to ensure its continued integrity while en route and to maintain periodic contact with the transport vehicle, at a frequency not to exceed 2 hours.

5. SHIPMENTS BY U.S. WATERS

This chapter provides guidance on the U.S. Nuclear Regulatory Commission (NRC) requirements specific to physical protection of spent nuclear fuel (SNF) shipments that travel on U.S. waters. Requirements discussed in this chapter are in addition to and not a substitute for requirements listed elsewhere in this NUREG that are applicable to physical protection of all SNF shipments regardless of the mode of transport (road, rail or water).

For the purposes of these requirements, U.S. waters are considered to extend 3 nautical miles from U.S. land territory, with the exception of small offshore islets. U.S. land territory includes the 48 contiguous States, Alaska, the eight largest islands of Hawaii, Puerto Rico, the Northern Mariana Islands, Guam, American Samoa, and the three major islands of the U.S. Virgin Islands. Security of sea shipments between 3 and 12 nautical miles out is the responsibility of the U.S. Coast Guard, which also publishes detailed security requirements pertaining to U.S. ports (Subchapter H, "Maritime Security" to Chapter I of 33 CFR). Information identified in 10 CFR 73.37(e)(4) shall be protected against unauthorized disclosure as specified in 10 CFR 73.21, "Protection of Safeguards Information: Performance Requirements," and 10 CFR 73.22, "Protection of Safeguards Information: Specific Requirements."

5.1. Protection of Waterborne Shipments

The purpose of the requirements in 10 CFR 73.37(e) is to ensure that the licensee has provided a capability for immediate, active protection of a waterborne shipment of SNF against theft, loss, diversion, or radiological sabotage while the shipment vessel is docked at a U.S. port, and for exports and imports, from the time the import enters the 3-mile zone until it arrives at a U.S. port, and from the time the export departs a U.S. port until it leaves the 3-mile zone. The licensee is to achieve this objective by ensuring that personnel who are on duty on or near the shipment vessel have the capability to delay or impede such acts and to request assistance promptly from local law enforcement agency (LLEA) response forces.

The requirements in 10 CFR 73.37(e) are applicable at any U.S. port through which the shipment may pass. If a U.S. port is used during a transport, the licensee shall coordinate with the U.S. Coast Guard and local port authorities to ensure that all parties are appropriately informed of the material being received at the port and to ensure that the physical protection requirements of 10 CFR 73.37, "Requirements for Physical Protection of Irradiated Reactor Fuel in Transit," are complied with during port operations. In addition to the provisions of 10 CFR 73.37(b), a shipment vessel, while docked at a U.S. port, must meet either of the following escort requirements:

- two armed escorts stationed on board the shipment vessel, or stationed on the dock at a location that will permit observation of the shipment vessel or the points of access to the SNF shipment, if it is located in an enclosed cargo compartment

- a member of an LLEA, equipped with normal LLEA radio communications, who is stationed on board the shipment vessel, or on the dock at a location that will permit observation of the shipment vessel or the points of access to the SNF shipment, if it is located in an enclosed cargo compartment

Section 73.37(e)(2) requires as permitted by law, that each armed escort carry a minimum of two weapons. Each weapon should provide separate and distinct response capabilities (e.g., a handgun and a rifle or shotgun). This requirement does not apply to LLEA personnel performing escort duties.

Additionally, 10 CFR 73.37(e)(3) requires that while a shipment vessel is within U.S. territorial waters, an individual must accompany it who will ensure that the shipment is unloaded only as authorized by the licensee. This requirement ensures that the SNF shipment is under surveillance against possible theft, loss, diversion, or radiological sabotage and that any necessary request for assistance from LLEA response forces is communicated promptly.

The individual accompanying the shipment may be provided in a number of ways. The individual may come aboard the vessel at the time of loading (or stopover) in a foreign port, may come aboard as the ship nears U.S. territorial waters (as do ship's pilots), or may be a ship's officer who assumes this surveillance role as the SNF shipment enters U.S. waters.

The accompanying individual should be familiar with the vessel's itinerary. Before the vessel enters port, the individual should verify that the SNF shipment is intact and has not been tampered with. During unloading of any cargo in U.S. territory, the individual should ensure that the SNF shipment is unloaded only at the port that the shipper (licensee) has authorized. Should any deviation from authorized handling of the SNF shipment occur, the individual should bring the matter to the immediate attention of the ship's most senior officer present. If it appears that the SNF shipment is likely to be unloaded at a port other than the planned destination, the individual should also immediately notify the NRC, the U.S. Coast Guard, and the licensee by ship-to-shore radio, radiotelephone, or other available means.

5.2.   Communications for Waterborne Shipments

Section 73.37(e)(4) requires each armed escort to be equipped with redundant communications capabilities that provide for two-way communication among the escorts, vessel, movement control center, LLEA, and one another at all times. The purpose of the communications requirements for waterborne shipments is similar to that of road and rail shipments, which is primarily to ensure that a capability exists to contact the LLEA in case a threat is detected to the safety or security of the shipment. The communications equipment provided may be the ship's regular ship-to-shore communications device, but alternate communication methods should not be subject to the same failure modes as the primary communication method.

6. BACKGROUND INVESTIGATIONS FOR UNESCORTED ACCESS TO SPENT NUCLEAR FUEL IN TRANSIT

This chapter provides guidance on the U.S. Nuclear Regulatory Commission (NRC) requirements for granting and controlling unescorted access to shipments of spent nuclear fuel (SNF).

Before granting an individual unescorted access to SNF in transit, the licensee shall complete a background investigation of the individual, in accordance with 10 CFR 73.38, "Personnel Access Authorization Requirements for Irradiated Reactor Fuel in Transit," to determine if the individual is trustworthy and reliable. The requirements of 10 CFR 73.38 apply to vehicle operators, escorts, and any other individuals accompanying the SNF shipment during transport, as well as to movement control center personnel, reviewing officials, background screeners, and any other access authorization program personnel.

The licensee is solely responsible for granting and controlling unescorted access to SNF in transit. The licensee should use any information obtained as part of a criminal history records check solely for the purpose of determining an individual's suitability for unescorted access to SNF in transit.
Section 73.38(j) requires licensees, contractors, and vendors to develop, implement, and maintain written procedures for conducting background investigations for individuals applying for unescorted access to SNF or for reinstatement of unescorted access, and for individuals denied unescorted access.

The scope of the background investigation must include the previous 10 year period, or if 10 years of information is not available, then as many years in the past in which information is available, as required by 10 CFR 73.38(d).

6.1. Informed Consent

Before initiating an individual's background investigation for unescorted access, the licensee must inform the individual and obtain the individual's written consent in accordance with 10 CFR 73.38(d)(1). The individual shall be informed of the types of records that may be produced and retained, where these records are normally maintained, and the duration of records retention. The individual also shall be informed of his or her right to review the information and to ensure its accuracy and completeness, as well as to whom and under what circumstances the information may be released.

An individual may withdraw consent to a background investigation at any time. When an individual withdraws consent, all processing must cease as soon as practical. Withdrawal of consent is the same as withdrawal of the application for unescorted access.

6.2. Protection of Information

In accordance with the requirements of 10 CFR 73.38(f), licensees are required to protect information collected while performing a criminal history records check and background investigation. The following are guidelines for the protection of information:

- Each licensee that obtains an individual's criminal history records check, in accordance with 10 CFR 73.38(c)(1), shall establish and maintain a system of files and procedures for protecting the record and the personal information from unauthorized disclosure.

- As specified in 10 CFR 73.38(f)(2), the licensee may not disclose the record or personal information collected and maintained to persons other than the subject individual, his or her representative, or those who have a need to access the information in performing assigned duties in the process of determining access to SNF in transit. No individual authorized to have access to the information may disseminate the information to any other individual who does not have a need to know.

- The personal information obtained about an individual from a criminal history records check and background investigation may be transferred to another licensee in accordance with 10 CFR 73.38(f)(3). This transfer may only take place when both of the following conditions are met:

  – if the licensee holding the criminal history records check receives the individual's written request to disseminate the information contained in his or her file and

  – the acquiring licensee verifies information such as the individual's name, date of birth, Social Security number, sex, and other applicable physical characteristics for identification purposes

- The licensee shall make criminal history records checks obtained under this section available for examination by an authorized representative of the NRC to determine compliance with the regulations and laws as specified in 10 CFR 73.38(f)(4).

- Under 10 CFR 73.38(f)(5), the licensee shall retain all fingerprint and criminal history records checks received from the Federal Bureau of Investigations (FBI) (including data indicating no record), or a copy of such records if the individual's file has been transferred, for 5 years after termination of employment or 5 years after the individual no longer requires unescorted access to SNF in transit.

- After the required 5-year period, all personal information should be destroyed by a method that will prevent reconstruction of the information in whole or in part.

6.3.   Background Investigation Elements

6.3.1. Fingerprinting and FBI Identification and Criminal History Records Check

Each individual who is seeking or permitted unescorted access to SNF in transit is subject to fingerprinting and an FBI identification and criminal history records check as specified in 10 CFR 73.57, "Requirements for Criminal History Records Checks of Individuals Granted Unescorted Access to a Nuclear Power Facility or Access to Safeguards Information." The licensee shall review and use the information received from the FBI in the determination to grant or deny the individual unescorted access to SNF in transit.

As specified in 10 CFR 73.38(c)(2), fingerprinting and the identification and criminal history records checks and other elements of the background investigation are not required for the following individuals before granting unescorted access to SNF in transit:

- Persons identified in 10 CFR 73.59, "Relief from Fingerprinting, Identification and Criminal History Records Checks and Other Elements of Background Checks for Designated Categories of Individuals," and 10 CFR 73.61, "Relief from Fingerprinting and Criminal History Records Check for Designated Categories of Individuals Permitted Unescorted Access to Certain Radioactive Materials or Other Property,"

- Federal, State, and local officials, including inspectors, whose occupational status is consistent with the promotion of common defense and security or the protection of public health and safety relative to SNF in transit,

- Emergency response personnel who are responding to an emergency,

- An individual who has had a favorably adjudicated U.S. Government criminal history records check within the last 5 years, under a comparable U.S. Government program involving fingerprinting and an FBI identification and criminal history records check (e.g., National Agency Check, Transportation Worker Identification Credentials under 49 CFR 1572, Bureau of Alcohol Tobacco Firearms and Explosives background check and clearances under 27 CFR 555, Health and Human Services security risk assessments for possession and use of select agents and toxins under 42 CFR 73, hazardous material security threat assessment for hazardous material endorsement to commercial driver's license under 49 CFR 1572, Customs and Border Patrol's Free and Secure Trade Program) provided that he or she makes available the appropriate documentation. Written confirmation from the agency or employer that granted the Federal security clearance or reviewed the criminal history records check must be provided to the licensee. The licensee shall retain this documentation for a period of 3 years from the date the individual no longer requires access authorization relative to SNF in transit, and

- Any individual who has an active Federal security clearance, provided that he or she makes available the appropriate documentation substantiating that clearance. Written confirmation from the agency or employer that granted the Federal security clearance or reviewed the criminal history records check must be provided to the licensee. The licensee shall retain this documentation for a period of 3 years from the date the individual no longer requires access authorization relative to SNF in transit.

As mentioned previously, fingerprinting and the other investigative elements in 10 CFR 73.38 do not apply to Federal, State, local, or Tribal law enforcement personnel.

Section 73.38(d)(1) requires the licensee to notify each affected individual that the fingerprints will be used to secure his or her criminal history records check and shall inform the individual of the procedures for revising the record or including an explanation in the record, as specified in Section 6.7 of this guidance.

All fingerprints the licensee obtained under 10 CFR 73.38(d) must be submitted to the NRC for transmission to the FBI.

6.3.2. Collection of Fingerprints

The licensee should have an authorized official, such as a representative from a local law enforcement agency (LLEA), take the individual's fingerprints. However, an authorized official could be available through private entities, contractors, or an established onsite fingerprinting program. If the licensee has fingerprints taken at a facility other than that of a recognized Federal, State, or LLEA, then the licensee should ensure that the prints are taken legibly and match the identity of the individual named on the fingerprint card. In these cases, the individual taking fingerprints should, at a minimum, do the following:

- Be trained to take fingerprints (training to take fingerprints is offered through the FBI, or may be available from LLEAs and some professional associations).

- Verify the identity of the individual being fingerprinted by checking a Government-issued picture identification (e.g., passport or driver's license) and comparing the name on the fingerprint card to the name listed on the Government-issued identification.

- Sign the block on the fingerprint card labeled "SIGNATURE OF OFFICIAL TAKING THE FINGERPRINTS."

Section 73.38(d)(3) requires the licensee to submit one completed, legible standard fingerprint card (Form FD 258, ORIMDNRCOOOZ) or, where practicable, other fingerprint records for each individual seeking unescorted access to SNF in transit, to the NRC's Division of Facilities and Security, ATTN: Criminal History Program, Mail Stop T 06E46, Rockville, MD 20852. Overnight mail is preferred. Include with the cards the name and address of the person to whom the individual's criminal history records check should be returned to for review.

Copies of the required forms may be obtained by writing to the Office of Information Services, U.S. Nuclear Regulatory Commission, Washington, DC 20555, by calling 301-415-7232, or by e-mail to forms.resource@nrc.gov. Alternative formats for requesting the forms are set forth in 10 CFR 73.4, "Communications." The licensee shall establish procedures to ensure that the quality of the fingerprint impressions results in a low rejection rate because of illegible or incomplete cards.

Licensees also may submit fingerprints to the NRC electronically. However, for many licensees this option may be prohibitive because of the cost associated with purchasing electronic fingerprinting equipment. To establish an electronic fingerprinting program with the NRC, contact the NRC's Facility Security Branch at 301-415-6511. Electronic submission of fingerprints to the NRC must come directly from the licensee.

Once the NRC has reviewed submitted fingerprint cards for completeness, they will be scanned and transmitted electronically to the FBI. The cards are retained and secured for approximately 1 month, after which time they are destroyed in accordance with Federal guidelines.

Any Form FD 258 fingerprint record containing omissions or evident errors will be returned to the licensee for corrections. The fee for processing fingerprint checks includes one resubmission if the FBI returns the initial submission because the fingerprint impressions are unclassifiable (i.e., cannot be read by the FBI and used to identify the individual). The one free resubmission must include the FBI transaction control number (TCN). Additional guidance on resubmittal of fingerprint cards can be found in Section 6.3.3.

Fingerprints may be unclassifiable for several reasons:

- incomplete impressions (fingers not completely rolled from one side of the nail to the other)

- left and right hands reversed on the fingerprint card

- the same hand or finger printed twice on the card

- fingerprints are not clear and distinct (smudged, uneven, too dark or light)

- fingerprints on card missing or partially missing without an explanation

To avoid FBI rejection of fingerprints as unclassifiable, the licensee must ensure they are of good quality and do not include any of these deficiencies, and that instructions on the back of the fingerprint card have been followed completely.

The FBI has provided guidance on taking fingerprints for submission to the FBI at http://www.fbi.gov/hq/cjisd/takingfps.html.

Fees for processing fingerprint checks are due upon application. Licensees shall submit payment with the application for processing fingerprints by corporate check, certified check, cashier's check, or money order made payable to "U.S. NRC." (For guidance on making electronic payments, contact the Facilities Security Branch, Division of Facilities and Security, at 301-415-7404.) Combined payment for multiple applications is acceptable. The application fee is the sum of the user fee that the FBI charged for each fingerprint card or other fingerprint record the NRC submitted on behalf of the licensee plus an NRC processing fee, which covers administrative costs associated with NRC handling of the licensee's fingerprint submissions. The NRC publishes the amount of the fingerprint application fee on the NRC public Web site. Once the NRC receives the FBI identification and criminal history records check results, it will forward them to the licensee.

6.3.3. Resubmittal of Fingerprints

The overwhelming majority of fingerprint cards are returned as classifiable (i.e., can be read by the FBI and used to identify the individual). However, if the FBI returns the initial fingerprint submission because the fingerprint impressions cannot be classified, the licensee may retake and resubmit them (i.e., through a new Form 258 or electronic submission) for a second attempt. The licensee will not be charged for the resubmission if the licensee provides a copy of the FBI response indicating that the fingerprints could not be classified or if they provide the FBI TCN.

If the FBI is unable to classify the second submission of fingerprints, the licensee may submit additional fingerprint impressions as follows:

- The third fingerprint card submission will require payment of an additional processing fee.

- If the third submission is also returned as unclassifiable, the licensee may submit a fourth set of fingerprints. An additional fee is not required because the fee for the third submission includes one resubmission. As with the second submission, the FBI's response or TCN must be included or the submission may be treated as a new submission and an additional fee may be imposed. The licensee may opt to submit the third and fourth sets of prints together to avoid a potential delay in response. If the third set is returned as unclassifiable, the NRC will automatically resubmit the fourth set.

- If the fourth submission is returned as unclassifiable, the licensee should submit six additional fingerprint cards to the NRC for the individual. All six cards will be forwarded to the FBI, which will select the best quality prints from each card to make a complete set of fingerprints. An additional processing fee is required, which covers the processing of all six fingerprint cards but does not include an additional resubmission.

- If the FBI is unable to obtain classifiable fingerprints from the six cards based on conditions other than poor quality (e.g., medical conditions or physical anomalies that prevent the taking of readable prints), then the NRC will automatically request a background check based on a name search for the individual and will forward the results to the licensee.

- No further submissions will be required and the licensee may consider the results of the name search-based FBI identification and criminal history records check as a component in determining the trustworthiness and reliability of the individual in accordance with 10 CFR 73.38(e).

6.3.4. Verification of True Identity

Licensees shall, as required by 10 CFR 73.38(d)(4), verify the true identity of an individual applying for unescorted access to SNF in transit to ensure that the individual is who he or she claims to be. The licensee shall review official identification documents (e.g., driver's license, passport, Immigration and Naturalization Service Form I-9 (if applicable), Government identification, birth certificate issued by a State, province, or country of birth) and compare the documents to personal information data the individual provided to identify any discrepancy in the information. The licensee should verify identity in the following ways:

- Compare a valid (not expired) official photo identification (e.g., driver's license, passport, Government identification) with physical characteristics of the individual.

- Cross-check and compare pertinent data from the background investigation with information supplied by the applicant (e.g., Social Security number, date of birth, physical characteristics of the individual).

Section 73.38(d)(4) requires the licensee to document the type, expiration, and identification number of the identification or maintain a photocopy of identifying documents on file under 10 CFR 73.38(c), certify and affirm in writing that the identification was properly reviewed and maintain the certification and all related documents for review during NRC inspections.

6.3.5. Employment History Evaluation

Section 73.38(d)(5) requires the licensee to document the employment history of the individual seeking unescorted access to SNF in transit. The evaluation shall consist of contacts with previous employers and should include the following information:

- verification of claimed periods of employment or unemployment of 30 days or more

- verification of periods of self-employment by reasonable methods (e.g., tax records, customers, coworkers)

- verification of length and nature of each period of employment, including reasons for termination and eligibility for rehire

- disciplinary history

- any other information that would adversely reflect on the trustworthiness and reliability of the individual as it relates to the individual being permitted unescorted access

- verification of activities during interruptions of employment in excess of 30 days

Section 73.38(d) requires the licensee to use their best effort to verify a 10-year employment history record; however, a minimum of a 3-year inclusive employment history immediately preceding the application for unescorted access is strongly recommended. Best efforts include such activities as documented attempts to contact previous employers and obtaining verification by telephone, letter, or other means. If the employment history check could not be accomplished for the entire 10-year period, the licensee should document this in an explanatory statement delineating the reasons. Section 6.3.11 below describes the steps a licensee should take if employment history cannot be readily verified.

For new hires without an employment history, the licensee will need to rely more on other aspects of the background investigation. A lack of employment history need not be a negative consideration in determining whether an individual is deemed to be trustworthy and reliable.

6.3.6. Verification of Education

Licensees must verify any claimed enrollment at an educational institution during the previous 10 years, as required by 10 CFR 73.38(d)(5). In addition, the licensee should verify the highest claimed post-high school attendance leading to a degree, regardless of when the degree was conferred. Section 6.3.11 below describes the steps a licensee should take when educational history cannot be readily verified.

6.3.7. Verification of Military Service

Section 73.38(d)(5) requires the licensee to verify any periods of active military service that an individual claims during the previous 10 years by reviewing a copy of the individual's DD Form 214 ("Certificate of Release or Discharge from Active Duty"). The licensee should also contact the last duty station, verify the dates of the individual's military service, determine the conditions under which the individual terminated his or her duty (i.e., honorable or dishonorable discharge), and if the individual would be eligible to serve again. The results of this contact should be documented. Section 6.3.11 below describes the steps a licensee should take when military service cannot be readily verified.

6.3.8. Credit History Evaluation

Section 73.38(d)(6) requires licensees to ensure that the full credit history of any individual who is applying for unescorted access to SNF in transit is evaluated. A full credit history evaluation must include, but is not limited to, an inquiry to detect potential fraud or misuse of Social Security numbers or other financial identifiers, and a review and evaluation of all of the information provided by a national credit-reporting agency on the individual's credit history. During the course of the credit history evaluation the licensee should ensure that its actions comply with all Federal and applicable State and local laws and regulations that govern credit history reports and access to such reports, including the Fair Credit Reporting Act.

For foreign nationals and U.S. citizens who have resided outside the United States and do not have an established credit history that covers at least the most recent 7 years in the United States, the licensee must document all attempts to obtain information about the individual's credit history and financial responsibility from some relevant entity located in the other country or countries.

6.3.9. Criminal History Review

Section 73.38(d)(7) requires the licensee's reviewing official to evaluate the results of the FBI criminal history records check for an individual applying for unescorted access to SNF in transit to determine whether the individual has a record of criminal activity that may compromise his or her trustworthiness and reliability. The scope of the licensee's review of the criminal history records check must cover all residences of record for the 10-year period preceding the date of application for unescorted access authorization. The licensee's reviewing official is required to evaluate all pertinent and available information in making a determination of unescorted access to SNF in transit. The reviewing official should consider the following:

- If negative information is discovered that the individual did not provide or that is different in any material aspect from the information provided by the individual, this information should be considered and decisions based on these findings must be documented.

- Any record containing a pattern of behaviors that indicates that the behaviors could be expected to recur or continue, or recent behaviors that cast doubt on whether an individual should have unescorted access to SNF in transit, should be carefully evaluated before any authorization of unescorted access to SNF in transit.

The licensee shall review the information received from the FBI and consider it, in conjunction with the trustworthy and reliability requirements included in Section 6.3 of this guide, in determining whether to grant the individual unescorted access to SNF in transit.

6.3.10. Character and Reputation Determination

The licensee must examine the character and reputation of an individual, as required by 10 CFR 73.38(d)(8), by conducting reference checks with current and former coworkers, neighbors, or friends. Reference checks may not be conducted with any person who is known to be a close member of the individual's family, including but not limited to the individual's spouse, parents, siblings, or children, or any individual who resides in the individual's permanent household. References used to verify periods of employment or unemployment also should not be used for the character and reputation determination.

The applicant's reputation for reliability and trustworthiness should be verified through contact with at least two references the applicant supplied and at least two additional references developed during the investigation. The reference's association with the applicant must be substantive, such that the licensee can make a meaningful determination of the individual's character and reputation. The determination should emphasize the following:

- criminal history

- illegal use or possession of a controlled substance

- abuse of alcohol, prescriptions, or over-the-counter drugs

- susceptibility to coercion or bribery

- any other conduct relating to an individual's trustworthiness and reliability to perform job duties

The licensee must document all contact with character references. In addition, the licensee must document the source of the developed references. Section 6.3.11 below describes what a licensee should do if references and developed references cannot be readily verified.

6.3.11. Inability or Refusal to Support a Background Investigation

As required by 10 CFR 73.38(d)(5), if a previous employer, educational institution, or any other entity with which the individual claims to have been engaged fails to provide information or indicates an inability or unwillingness to provide information within 3 business days of the request, the licensee shall do the following:

- Document the refusal, unwillingness, or inability in the record of investigation.

- Obtain a confirmation of employment, educational enrollment, and attendance, or other form of engagement claimed by the individual from at least one alternate source that has not been previously used.

Independent information may be obtained through interviews with anybody who knows or previously knew the individual, such as teachers, friends, coworkers, neighbors, and family members.

The NRC understands that simple verbal confirmations of past employment, education, etc., and timeframe may be all the information an employer or other confirming entity may be willing to provide on an individual. Although a simple confirmation of the nature and timeframe of past employment, education, etc., would not in itself suffice to permit a licensee to find an individual trustworthy and reliable, it would constitute independent corroboration of the accuracy of the individual's information about that period of personal history.

6.4.   Determination of Trustworthiness and Reliability

After obtaining and evaluating all of the information collected during the background investigation conducted in accordance with 10 CFR 73.38(d), the licensee shall determine whether the individual is trustworthy and reliable. If the licensee concludes that there is reasonable assurance that the individual is trustworthy and reliable and does not constitute an unreasonable risk for malevolent use of the material, then and only then may the licensee grant the individual unescorted access to SNF in transit. The licensee shall document its determination and the basis for it. The licensee shall maintain records of the determination for 5 years from the date the individual no longer requires access to SNF in transit.

The licensee is solely responsible for making a trustworthiness and reliability determination of an employee, contractor, or other individual who would be granted unescorted access to SNF while in transit. It is expected that licensees will use their best efforts to obtain the information required to conduct a background investigation to determine an individual's trustworthiness and reliability.

The background investigation consists of a series of data points. Lack of a single data point (e.g., inability to get employment history, inability to verify military record, inability to get additional references) should not necessarily result in a determination that the individual is not trustworthy and reliable.

6.5.   Reinvestigations

As required by 10 CFR 73.38(h), the licensee shall conduct fingerprinting, an FBI identification and criminal history records check, and a credit history reevaluation every 10 years for any individual who has been granted unescorted access to SNF in transit. The reinvestigations must be completed within 10 years of the date on which these elements were last completed and should address the 10 years following the previous investigation.

6.6.   Prohibitions

A licensee should not deny an individual unescorted access to SNF in transit solely on the basis of information received from the FBI involving an arrest more than 1 year old for which there is no information of the disposition of the case, or an arrest that resulted in dismissal of the charge or an acquittal.

Licensees should not use information received from a criminal history records check obtained under these requirements in a manner that would infringe on the rights of any individual under the First Amendment to the Constitution of the United States, nor shall licensees use the information in any way that would discriminate among individuals on the basis of race, color, religion, national origin, sex, age, disability, veteran status, or genetic information.

6.7.   Right to Correct and Complete Information

Section 73.38(k) requires before any final adverse determination that the licensee make available to the individual the contents of any criminal history records check obtained from the FBI for the purpose of ensuring correct and complete information. The licensee must maintain written confirmation by the individual of receipt of this notification for a period of 1 year from the date of the notification.

If, after reviewing his or her criminal history record, an individual believes that it is incorrect or incomplete and requests to change, correct, or explain anything in the record, the individual may initiate procedures to challenge the determination. These procedures would include challenging the criminal history records to the law enforcement agency that contributed the questioned information. Or, an individual may challenge the accuracy or completeness of any entry on the criminal history record by contacting the FBI at:

Federal Bureau of Investigation
Criminal Justice Information Services (CJIS) Division
ATTN: SCU, Mod. D-2
1000 Custer Hollow Road
Clarksburg, WV 26306

Instructions for a challenge to the FBI are set forth in 28 CFR 16.30, "Purpose and Scope," 28 CFR 16.31, "Definition of Identification Record," 28 CFR 16.32, "Procedure to Obtain an Identification Record," 28 CFR 16.33, "Fee for Production of Identification Record," and 28 CFR 16.34, "Procedure to Obtain Change, Correction or Updating of Identification Records." The FBI will forward the challenge to the agency that submitted the data, and will request that the agency verify or correct the challenged entry. Upon receipt of an official communication directly from the agency that contributed the original information, the FBI Identification Division makes any necessary change to the individual's criminal history record.

The licensee cannot make a final adverse determination based only upon the individual's criminal history record review until the licensee receives the FBI's confirmation or correction of the record.

Section 73.57(e) requires that the licensee provide at least 10 days for an individual to initiate action to challenge the results of an FBI criminal history records check after the record has been made available to the individual for review, although the licensee may allow more time. The licensee may use its own judgment for other elements of the background information.

APPENDIX A
SAMPLE TABLES

Table A-1 Sample Route Mileage Summary

STATE	WAYPOINT/INTERSECTION	ROAD	DIR	TOWARD	MILES TO NEXT WAYPOINT	CUMULATIVE ROUTE MILES
	DEPART ABC Licensee					
STATE MILEAGE SUBTOTAL						
STATE MILEAGE SUBTOTAL						
	ARRIVAL XYZ Licensee					
STATE MILEAGE SUBTOTAL						
ROUTE TOTAL						

A-1

Table A-2 Sample Detailed Route Plan[1]

State:

Statewide Emergency Response Number:

Governor's Designee and Phone Number:

Communications: Satellite:
                      Cell Coverage:

County/Reservation[2]	Contacts[3]	Safe Havens[4]	Routes

---

[1] The reviewer should verify that this portion of the application is designated correctly. The information associated with the location of safe havens is designated Safeguards Information.

[2] Identify each county and participating Tribe's reservation within the State through which the shipment will pass.

[3] Identify the 24-hour emergency contact information for each county, when practicable. Twenty-four hour numbers are preferred, but daytime phone numbers supplemented by after-hours phone numbers are acceptable.

[4] Identify the safe havens and their locations.

NRC FORM 335 (12-2010) NRCMD 3.7	U.S. NUCLEAR REGULATORY COMMISSION	1. REPORT NUMBER (Assigned by NRC, Add Vol., Supp., Rev., and Addendum Numbers, if any.)
	BIBLIOGRAPHIC DATA SHEET *(See instructions on the reverse)*	NUREG-0561 Rev. 2 FINAL

2. TITLE AND SUBTITLE

Physical Protection of Shipments of Irradiated Reactor Fuel

3. DATE REPORT PUBLISHED	
MONTH	YEAR
April	2013

4. FIN OR GRANT NUMBER

5. AUTHOR(S)

A.G. Garrett, S.L. Garrett, Pacific Northwest National Laboratory
A.S. Giantelli, R.C. Ragland, U.S. Nuclear Regulatory Commission

6. TYPE OF REPORT

Technical

7. PERIOD COVERED (Inclusive Dates)

8. PERFORMING ORGANIZATION - NAME AND ADDRESS (If NRC, provide Division, Office or Region, U. S. Nuclear Regulatory Commission, and mailing address; if contractor, provide name and mailing address.)

Pacific Northwest National Laboratory, P.O. Box 999, Richland, WA 99352
Division of Security Policy, Office of Nuclear Security and Incident Response
U.S. Nuclear Regulatory Commission
Washington, DC 20555-001

9. SPONSORING ORGANIZATION - NAME AND ADDRESS (If NRC, type "Same as above"; if contractor, provide NRC Division, Office or Region, U. S. Nuclear Regulatory Commission, and mailing address.)

Same as above

10. SUPPLEMENTARY NOTES

Supercedes NUREG-0561, Revision 1

11. ABSTRACT (200 words or less)

This guidance document sets forth means, methods and procedures that the staff of the U.S. Nuclear Regulatory Commission (NRC) considers acceptable for satisfying the requirements for the physical protection of spent nuclear fuel (SNF) during transportation by road, rail and water, and for satisfying the requirements for background investigations of individuals granted unescorted access to SNF during transportation. Chapter 1 discusses the regulatory basis and definitions applicable to shipments of SNF. Chapter 2 corresponds to the general requirements for the physical protection of SNF while in transit. These requirements apply to all SNF shipments regardless of the mode of transportation used for a particular shipment. Chapters 3, 4, and 5 discuss additional requirements specific to each particular transport mode—road, rail, or U.S. waters. Chapter 6 discusses the requirements for background investigations of individuals with unescorted access to SNF in transit.

12. KEY WORDS/DESCRIPTORS (List words or phrases that will assist researchers in locating the report.)

Transportation, security, security measures, licensee, protection, physical protection, shipment, shipping, import, export, irradiated, reactor, reactor fuel, nuclear fuel, spent nuclear fuel, SNF, uranium, plutonium, fission products, transit, transport, regulations, emergency response, safeguards, threat, sabotage, theft, diversion, response force, schedule, safe haven, telemetric monitoring, escort, armed, NRC, LLEA, local law enforcement, approval, route, route plan, advance notification, background investigation, fingerprints, unescorted access, credit history, FBI, criminal history, employment history, military service, reputation, trustworthiness, reliability, nuclear security, incident response, Nuclear Regulatory Commission, NUREG-0561

13. AVAILABILITY STATEMENT

unlimited

14. SECURITY CLASSIFICATION

(This Page)

unclassified

(This Report)

unclassified

15. NUMBER OF PAGES

16. PRICE

UNITED STATES
NUCLEAR REGULATORY COMMISSION
WASHINGTON, DC 20555-0001

OFFICIAL BUSINESS

NUREG-0561, Rev. 2
Final

Physical Protection of Shipments of Irradiated Reactor Fuel

April 2013